YOUR KNOWLEDGE HAS VALUE

- We will publish your bachelor's and
  master's thesis, essays and papers

- Your own eBook and book -
  sold worldwide in all relevant shops

- Earn money with each sale

Upload your text at www.GRIN.com
and publish for free

Ataliba Miguel

Gazelle Field Development

GRIN Verlag

Bibliografische Information der Deutschen Nationalbibliothek:

Die Deutsche Bibliothek verzeichnet diese Publikation in der Deutschen National-
bibliografie; detaillierte bibliografische Daten sind im Internet über http://dnb.d-
nb.de/ abrufbar.

Imprint:

Copyright © 2014 GRIN Verlag GmbH
Druck und Bindung: Books on Demand GmbH, Norderstedt Germany
ISBN: 978-3-656-74606-5

This book at GRIN:

http://www.grin.com/en/e-book/280478/gazelle-field-development

Gazelle Field Development

Prepared By:

Ataliba Miguel

August 2014

Word Count 2668

List of Contents

ABBREVIATIONS

BOPD	BARRELS OF OIL PER DAY
BS&W	BASIC SEDIMENT AND WATER
CAPEX	CAPITAL EXPENDITURE
FPSO	FLOATING PRODUCTION AND OFFLOADING
FSU	FLOATING STORAGE UNIT
GBS	GRAVITY BASED STRUCTURE
HP	HIGH PRESSURE
LP	LOW PRESSURE
MMbbl	MILLION BARRELS
OIW	OIL IN WATER
OPEX	OPERATING EXPENDITURE
TLP	TENSION LEG PLATFORM
STOIIP	STOCK TANK OIL INITIALLY IN PLACE
WHP	WELLHEAD PLATFORM
WI	WATER INJECTION

1. Introduction

The Gazelle field is located offshore Kuwait in Iranian waters very close to Shat Al Arab. The field discovered with the initial exploration well in 1990. Oil production is via a steel piled jacket platform, and the export system is via FSO to a shuttle tanker. After a number of years of production, the oil rates have come off plateau with increasing amount of water cut. Field sensitivities require a new development of the Gazelle field.

According to the project brief, three field developments options were proposed such as:

- Refurbishment and upgrade of present system
- Replacement of the oil export via the FSO and
- Complete system replacement

The approach used to evaluate and recommend the most suitable development concept was based on

- Economics (technical costs)
- Flexibility (project implementation) and
- Abandonment strategies

2. Key Characteristics Gazelle field

Gazelle field sensitivities outlined in the following sections provides a development background.

2.1 Field description

The field is located offshore in 45 km to the nearest landfall, which is a desert remote region with very limited infrastructure. The water depth is 100 meters.

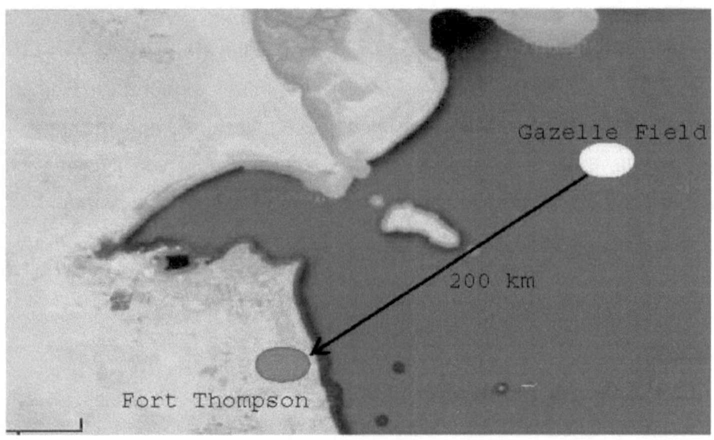

Figure 1 – Gazelle Field location

The Gazelle field produces from two reservoirs – the Gazelle Upper and Gazelle Deep. For further details, please see the Table 1.

Table 1 – Petroleum engineering

Field Description	Gazelle Upper	Gazelle Deep
Reservoir Depth	10,000 ft	12,000 ft
Oil water contact	10,200 ft	12,400 ft
Bubble point	3,800 psia	3,800 psia
Initial GoR	450 scf/bbl	500 scf/bbl
Initial Pressure	5000 psia	6100 psia
Light sweet crude	36° API	36° API
Expectation STOOIP	300 mmbbl	450 mmbbl

A pressure support will be required to enhance the recoverable reserves from the Gazelle Deep reservoir. Additional water injection wells are required in the lower bearing zone for pressure support.

2.2 Gazelle Field Information

Gazelle production is via steel-jacketed platform in 100 meters water depth. Oil export is via FSU to a shuttle tanker. The field has been producing over 6 years and has come off plateau. The current production rates are around 75,000 bpd with water cut 15%. The process configuration in the Gazelle steel-jacketed platform consists of:

> ➤ Two 50% trains each two-stages of two-phase separation and a surge vessel to stabilise the crude oil to FSU and tanker specifications.
> ➤ Gas from HP separator and LP separators is compressed and conditioned for use as fuel gas. Gas from surge vessel plus any excess is flared.

The deck loadings and space availability at the Gazelle steel jacketed platform are very limited.

3. Development considerations

3.1 System upgrade

The current separation system cannot handle the changes in the well stream, which uses gravity settling in the FSU tanks. Therefore, retrofitting the existing separation system in the Gazelle platform during the FSU refurbishment process technically proves to be a viable option.
The plan shall include the following:

> ➤ Upgrade of separation system
> ➤ FSU refurbishment
> ➤ Installation of new jack up platform (wellhead + water treatment/injection modules + gas injection and drilling module)

3.2 Assumptions

The integrated well caisson in the jack up shall contain slots for ten wells, risers and J-tubes [4]. The drilled platform wells shall be completed using a cantilever jack up drilling rig [4]. Workover on platform wells are of easy access. The drilling methodology used will follow the same trend for all field development options proposed.

3.2 Separation process

The retrofit in the separation process consists in changing the existing separation trains from two-phase system to three-phase system; this option technically proves to be the best to handle initially the significant gross water cut at Gazelle platform. The gross liquids from three-stage separation are routed to the new jack up wellhead platform for further treatment. The platform is equipped with crude oil heat treaters (Tubular heaters), free water knockout drums (FWKO) and degassers.

3.4 Flow scheme

From wellhead to point of export, oil routes on a three-phase two-stage train separation system. The initial gross separation occurs in the HP separator, where gas is flashed to obtain a maximum liquid recovery. From HP separator the oil flows to LP separator and to oil surge vessel. Oil from surge vessel, which operates as a three-phase separator, flows to free water knockout drum (FWKO) for further water removal before passing through heater treaters and degasser vessel for final storage and offload via shuttle tanker.

Produced water from separator vessels routes to the water treatment unit (multiliner hydrocyclone vessel). The water is batch treated to meet the injection/disposal requirements. The oil rejected is recycled back to low-pressure separator. Degassing vessel is installed downstream of the multiliner hydrocyclone.

The gas out from separator vessels commingled passes through a scrubber before entering gas compressors. Compressed gas passes through a dehydration process before using for fuel and future re-injection purposes.

The economics associated to gas export are not feasible. The CAPEX to commission a gas export pipeline proves to be not very attractive worth to invest.

Figure – 2: PFD – Refurbish & upgrade

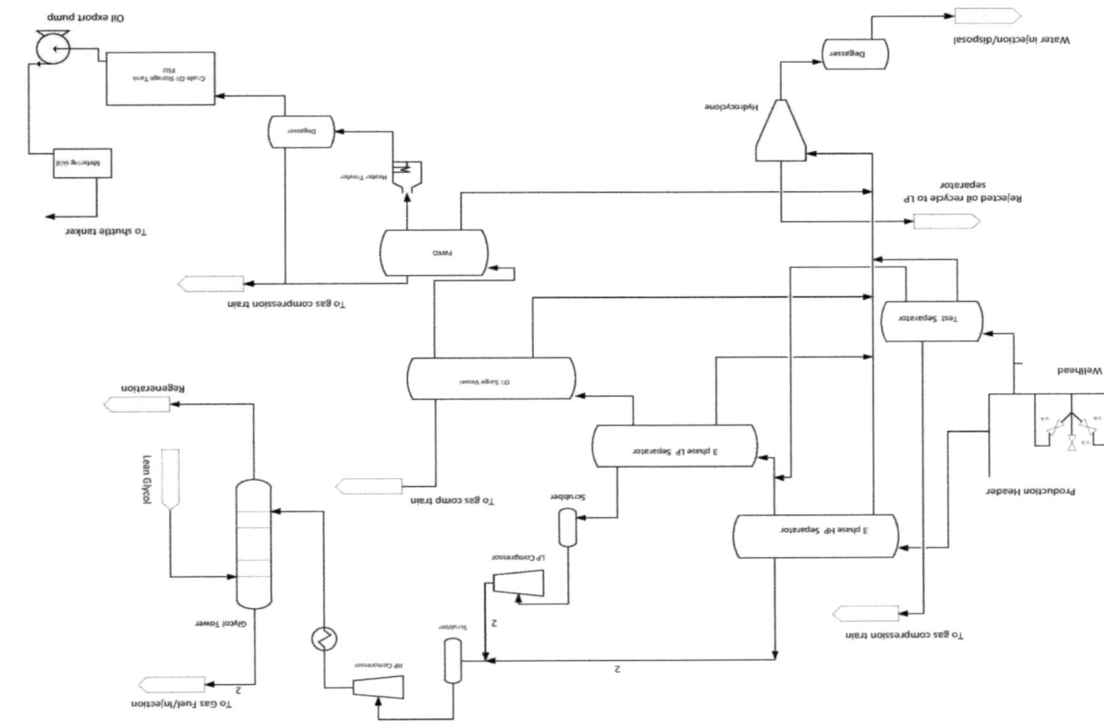

3.5 Platform replacement (pipeline)

This development option will consist of:

> ➢ New central processing platform (concrete)
> ➢ New wellhead platform with drilling rig and
> ➢ Oil export subsea pipeline from central processing platform

The following assumptions for this option includes the removal of existing Gazelle steel piled platform, and installation of a concrete gravity base CPP bridge linked to a wellhead platform with drilling capabilities.

From wellhead to point of export, the separation process will be via a three-phase three-stage system. The initial gross separation occurs in the HP separator. Downstream to HP separator a MP and LP separator increases the yield of liquid recovery. Oil leaving LP separator flows to an oil treater unit to stabilize the oil for storage in the base gravity tanks. The export of stabilized oil is via subsea pipeline.

Produced water from HP, MP and LP separators will be batch treated in the water treatment module prior to injection/disposal via wellhead platform. Conditioned gas from HP compressor is to use for fuel utilities purposes. Any excess gas is re-injected via dedicated gas disposal/lift wells.

Figure – 3: PFD – Replace Gazelle and pipeline

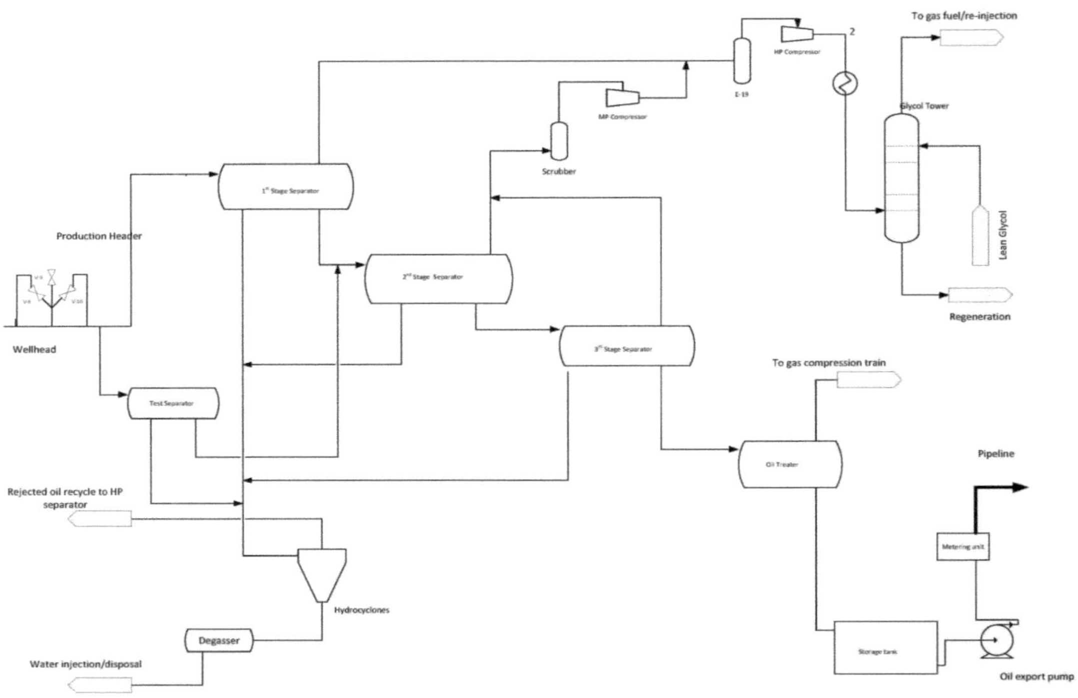

3.6 Complete System Replacement

The development option will consist of:

> ➢ New central processing platform (piled)
>
> ➢ New wellhead platform (jack up) with drilling rig and
>
> ➢ FSU replacement and oil export via dedicated shuttle tanker

This option used the same process configuration of the previous development option, the only difference being that the oil exported is via a new FSU to a dedicated shuttle tanker.

Figure – 4: PFD – Complete system replacement (no pipeline)

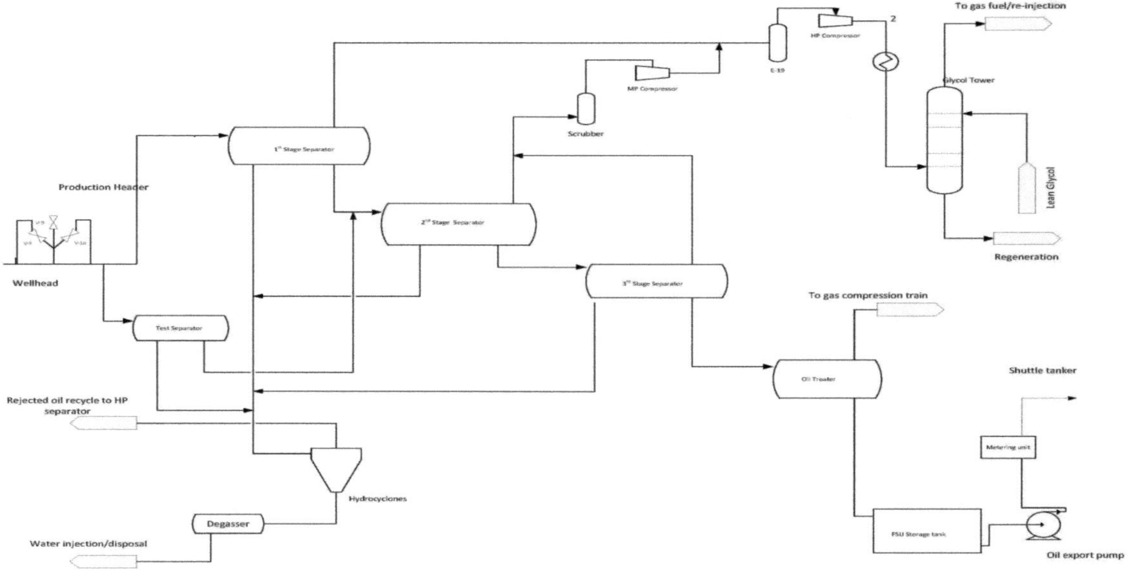

4. Flow Assurance

The major concern of flow assurance likely encountered in the proposed schemes is the high increase in water cut. As the reservoir depletes and water breaks through, several problems such as, corrosion, scale tendency, hydrates and sand production impairs the normal operation of the processing system.

Increased water cut in Gazelle field provides stability for crude oil emulsions. Water oil emulsion leads to rapid agglomeration of hydrate formation in flowlines [9]. Shutdown and restart conditions at Gazelle will occur during FSU refurbish and separation system upgrade; therefore, there is a window for hydrate formation during transient operations (i.e. shutdown and restart conditions due to low temperatures).

The injection system oftentimes is a combination of formation-produced water and seawater, the incompatible mixing of both waters yields precipitation and scaling tendency of scale salts and solids [2]. Scales form precipitates when pressure and temperatures decreases during production. The noticed effects are in clogged flowlines, forming oily sludge and forming emulsions difficult to break [10].

Corrosion triggers in the presence of various dissolved gases encountered in produced water. Bacterial action oftentimes causes H_2S corrosion by sulphate reducing bacteria.

As reservoir, pressure falls down and the field increases in water cut, sand and fine particles starts to build up. The adverse effect of sand build occurs at the bottom of separators, reducing noticeably the separators removal efficiency. Sand causes severe erosion/corrosion problems.

Produced water has higher concentrations of metals such as iron; when iron reacts with oxygen in the air, produces certain solid particles that interfere with hydrocyclones operation oftentimes leading to plug formation during water injection process.

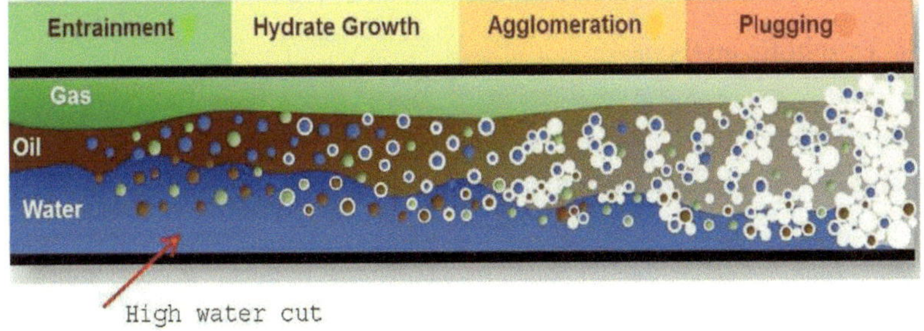

Figure – 5: Schematic of hydrate formation in high water cut systems (adapted from Joshi et al, 2013 [1])

The proposed solutions to prevent flow assurance issues are summarized in the table as shown below.

Table – 2: Flow assurance solutions

	Direct Prevention	**Comments**
Hydrates	Use LDHI's such as kinetic inhibitors or anti agglomerants upstream wellhead in downhole	• More cheaper compared to traditional methanol inhibition and more ecofriendly
Scale	Scale inhibitors injected downhole	• Needs compatibility with sea and formation water. • Balanced adsorption-desorption properties. • Low toxicity and low cost
Sand	Jet sand washers and desand slurry drain	• Installed in separator vessels

5. Economic analysis

Per barrel cost was the methodology adopted to evaluate the economic sensitivities of the proposed development options. It is very useful method to evaluate the cost of an existing producing field when there are some production constraints (such as Gazelle with high water cut) or when making technical comparisons between projects in the same geographical area.

$$\text{Per barrel cost} = (\text{CAPEX} + \text{OPEX}) / \text{Cum Production}$$

5.1 OPEX

The fixed OPEX is proportional to the capital cost of the items operated, therefore is a percentage of cumulative CAPEX.

The variable OPEX is proportional to the throughput and it relates to the production rate.

5.2 CAPEX

The consideration given in spending the CAPEX was based on several sensitivity runs carried out on the economic analysis spreadsheet provided. The rationale adopted was to minimize the gap between the CAPEX and revenue production [7]. Typically spreading the CAPEX across as many years as possible and start spending early allows enough time for construction prior the scheduled date of installation. An assumption during the period of commissioning and installation is that the gross production will cease to allow for the refurbishment/upgrade/replacement operations; therefore, for a less impact in the economics (cashflow) it is always preferable to spend the CAPEX when there is income revenue from gross production.

Full detailed economic analysis is available in the appendices for each of the options proposed. The following sections summarize the technical costs of each development scenario.

5.3   Cost – option one

The base case adopted in this analysis took into consideration the following assumptions:

- Using existing Gazelle steel piled
- Wells CAPEX spread across 2 years starting at year 4. CAPEX per well $M25 (10 wells to be drilled and completed).
- New jack up wellhead platform comprising 3 modules (drilling, water treatment and injection). The cost breakdown as follows:
  - Jack up = $M70
  - Drilling = $M75
  - Injection = $M98
  - Treatment = $M100
- The retrofit of separation system = $M125

Table – 3: Cost summary

Refurbish				
CAPEX ($mm)	Wells	Jack up wellhead	FSU refurb	Separation Upgrade
	250	343	100	125

Cum oil	Cum Capex	Cum Opex	$/bbl
253	818	1905	10.494

5.4 Cost – option two

The base case adopted in this analysis took into consideration the following assumptions:

- Gazelle replacement with a pipeline
- The wells CAPEX spread across 2 years starting at year 6. CAPEX per well $M25 (10 wells to be drilled and completed)

- New platforms CAPEX $M763, the cost breakdown as follows:
    - Jackets = $M(70+150) (jack up + concrete)
    - Drilling = $M75
    - Injection = $M98
    - Treatment = $M100
    - Production = $M125
    - Accommodation = $80
    - Utilities = $65
- Commissioning a new export 200 km pipeline CAPEX $M450.

Table – 3: Cost summary

Replace Gazelle + pipeline			
CAPEX ($mm)	Wells	Gazelle replacement	Pipeline
	250	763	450

Cum oil	Cum Capex	Cum Opex	$/bbl
266	1238	1398	11.8

5.5   Cost – option three

The base case adopted in this analysis took into consideration the following assumptions:

- Complete system replacement
- The wells CAPEX spread across three even years starting at year 2. CAPEX per well $M25 (10 wells to be drilled and completed)
- New platforms CAPEX $M738, the cost breakdown as follows:

  - Jackets = $M(70+125) (jack up + piled)
  - Drilling = $M75
  - Injection = $M98
  - Treatment = $M100
  - Production = $M125
  - Accommodation = $80
  - Utilities = $65

- Floating storage unit replacement CAPEX $M150

Table – 4: Cost summary

Complete system replace			
CAPEX ($mm)	Wells	Gazelle replacement	FSU replace
	250	738	150

Cum oil	Cum Capex	Cum Opex	$/bbl
266	1013	2065	12.4

6 Development choice

When evaluating the choice for the jackets six concepts were analysed as follows [8]:

- Jack-up construct
- Steel piled
- Concrete
- Semisubmersible
- TLP and
- SBM

The screening stage used to evaluate the concepts was through a combination of qualitative criteria to satisfy technical implementation, with cost versus benefit of the chosen concept.

In this development study, the concepts, which utilize subsea wells, were not subject to consideration during the screening evaluation. Out of the six concepts, Semisubmersible, SBM and FPSO are the ones that require subsea development completion.

The reasons as outlined below provide justification:

- Gazelle field water depth does not justify the need to develop the field using subsea completions
- The subsea wells CAPEX drives high the cost per barrel ($/bbl) to develop the field
- Subsea wells development requires an additional MODU, increasing further the development cost [3]
- Subsea solutions only selected if platform-based development is not profitable [16]
- Potentially long tie flowlines [3] and the threshold for new wells and workover is too high [16]

6.1   Risk assessment

The risk assessment conducted was to evaluate the risk ranking based upon the type and nature of the offshore structures, installation hazards and geotechnical failures, which may cause serious consequences to health and environment [11].

The tables as shown below summarize the key points with overall risk ratings.

Table – 8: Risk assessment

Structure	Foundation	Safety Consequences (High/medium/low)	Colour code
Jack up / Piled / GBS	Fixed	High	
TLP	Floating attached by tension legs	Medium	
Semi-submersible	Floating attached by tension anchors	Medium/Low	
FPSO	Floating anchored	Low	

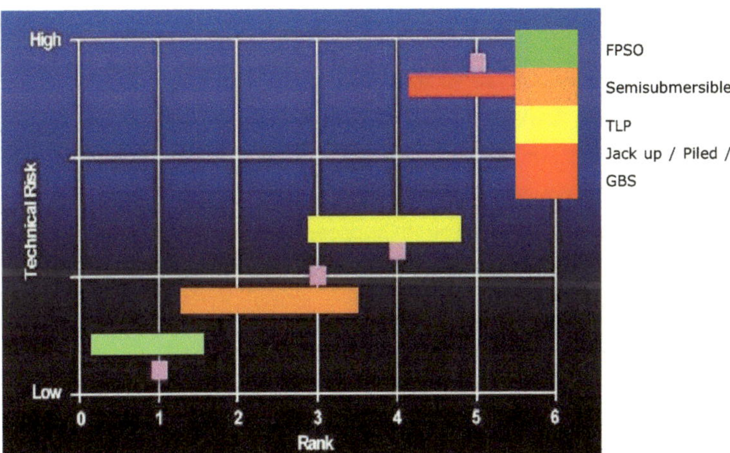

Figure – 5: Risk ranking (adapted from Alma Field Development [9]: Husky Oil)

7 Development analysis

The evaluation based on field sensitivities and screening analysis recommends developing the Gazelle field by using the existing Gazelle steel piled platform bridge linked to a new jack up wellhead platform with drilling capabilities and FSU (refer Figure 6).

The economics driven by technical cost per barrel is more attractive as compared with other development concepts. Refurbish and upgrade the present system increases the schedule flexibility, provides less field-commissioning trials (existing platform already in operation), is relatively inexpensive and eliminates the need of an early removal of the existing Gazelle steel piled platform; the abandonment expenditure (ABEX) postpones to decommissioning stage.

Platforms with drilling facilities have a substantially higher recovery factor

7.1. Comparisons

The choice forward on system replacement (including pipeline or FSU) for GBS (concrete) ranks low in terms of deliverability and schedule flexibility; it is not economically viable considering the cumulative oil reserves and presents problems in decommissioning and abandonment due to inability to relocate for another service (please refer table 5:appendix III) [5] [6].

The choice forward on semisubmersibles proves a relatively flexible schedule with few hook/installation manhours as the facility is built as a full-integrated system, however, the cost expenditure based on the cumulative oil it proves unattractive. A separate cost to operate and using drilling capabilities through subsea BOP and makes it even a disfavouring alternative selection (please refer table 7: appendix III).

TLP has a flexible construction schedule and makes it possible to operate the wells from the topside deck allowing workover interventions, but the wells Capex for TLP's drives high the technical costs to re-develop the field (please refer table 6:appendix III).

7.2. Technical risks

The technical risks presented in the table 8 above demonstrates that jack up platforms represent a high risk rank 5 as compared to TLP and Semisubmersible, but, it is worth to point out that this evaluation used historical industry statistics. Proper measures to mitigate and prevent hazards during installation of jack ups, provides a reasonable flexibility during installation.

7.3. Advantages

The advantages of using jack up platform with existing Gazelle apart from low cost and schedule efficiency are [16]:

- Jack up wellhead allows direct access for well workover.
- Flexible drainage strategy
- Technical flexibility
- Financial flexibility
- Lower operational risk
- Greater regularity
- The jack up platform will benefit from the existing Gazelle accommodation and utilities module

7.4. Disadvantages

The disadvantages of other two development concepts are as follow:
- Development cost per barrel proves not very attractive
- When installing the platform in the existing production well slots, there is a possibility of a deviation in pre-drilled production wells
- Possible operational issues during start up trials

Figure – 6: Recommended selection

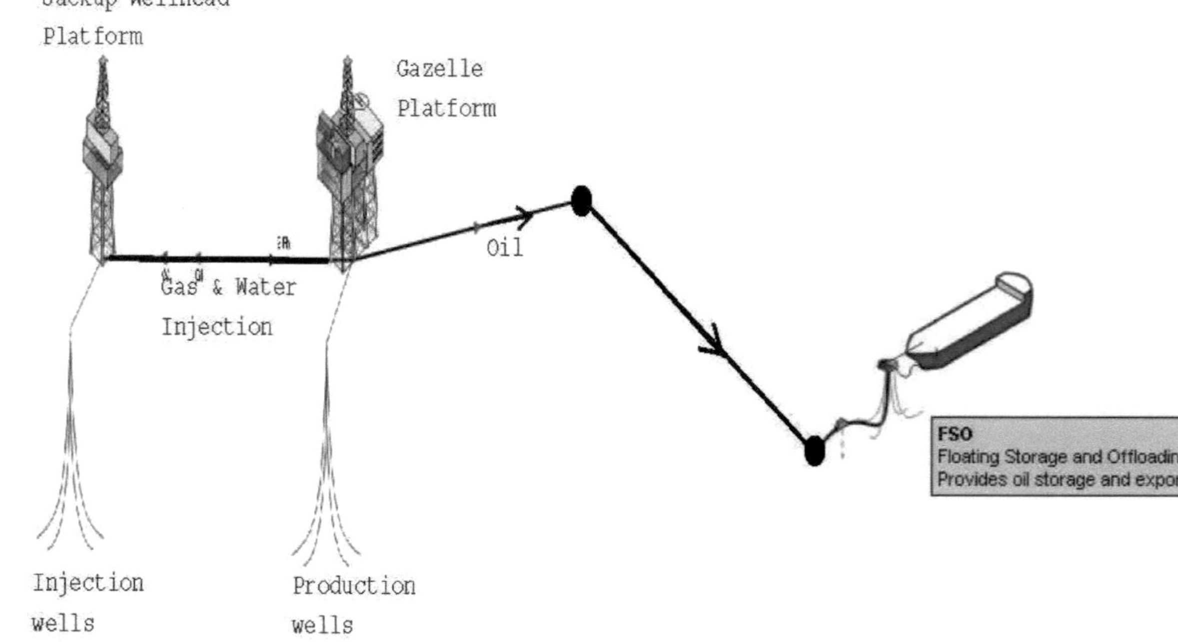

8 Decommissioning plan

The regional convention to be used is the Kuwait protocol for the Persian Gulf. This regional convention controls the pollution and requires removal of offshore structures. Therefore, the topsides of Gazelle structures are to be removed and brought to shore for disposal or reuse.

8.1 Decommissioning technology

The safety and environmental criteria along with the capability of technology is a key major factor influencing the decommissioning option. The most important technological areas widely used involve:

- ➢ Cutting offshore and subsea
- ➢ Offshore lifting and
- ➢ Onshore recovery and disposal

8.2 Decommissioning Process

The decommissioning process will include the following:

- ➢ Planning
- ➢ Commence consultation, submit programmes and programmes approved
- ➢ Start to P&A the wells
- ➢ Cessation of production
- ➢ Facilities hydrocarbon free
- ➢ Modules de-coupled
- ➢ Topsides removed
- ➢ Jackets removed – seabed clear

8.3 Key parameters

The key parameters taken in consideration for the achievement of the decommissioning option shall include the following:

- ➢ Contractors, including management and operatives, should be involved at an early stage. Good communication is vital.
- ➢ Health and safety of personnel should be a priority during the decommissioning process.
- ➢ Preparation for decommissioning should begin whilst the platform is still in operation, to properly map and document the state of the

facilities. Otherwise, knowledge of the Gazelle platform may not be complete.

> ➢ A cutting plan should be developed and a variety of cutting tools should be made available.

Having considered the above decommissioning options, the complete removal of existing Gazelle platform proves to be technically more feasible. Other operators in the region have exploration licenses; therefore, it is reasonable to reuse certain modules of the new jack up platform e.g. drilling module.

References

1. Joshi et al. (2013). "Experimental flowloop investigations of gas hydrate formation in high water cut systems". *Chemical Engineering Science*, 97 (0), pp. 198-209.

2. Haghtalab, A., Kamali, M.J. and Shahrabadi, A., (2014). Prediction mineral scale formation in oil reservoirs during water injection. *Fluid Phase Equilibria,* 373(0), pp. 43-54.

3. Lim. F., (). Dry or Wet Trees in Deepwater Developments from a Riser System Perspective. 2H Offshore Engineering.

4. Mobile Offshore Production Units. [Online]. Available at:
http://www.google.co.uk/url?sa=t&rct=j&q=&esrc=s&frm=1&source=web&cd=9&cad=rja&uact=8&ved=0CFcQFjAI&url=http%3A%2F%2Fwww.gustomsc.com%2Findex.php%2Fzoo%2Fbrochures%2Fdoc_download%2F660-production-jack-ups&ei=5QzGU5fJFOjI8wH2iYDoBw&usg=AFQjCNGYwV1NyaQ8AW_UJ8nqy9kwzhErgA&sig2=SEXF90SK7FDPDA0USrTpkw&bvm=bv.71126742,d.b2U

5. Morten. B., Manschot. D. and Olsen. T., (1999) The Siri Production Jack-up. OTC. Houston, Texas 3-6 May. Houston: Offshore Technology Conference
Accessed on 10th July 2014

6. Offshore Structures. [Online]. Available at:
http://www.essie.ufl.edu/sheppard/OCE3016/Offshore%20Structures.pdf

7. Powell, D. 2014, email 25th April, drew.powell@gaffney-cline.com

8. Production and Export Systems. [Online]. Available at:
http://www.huskyenergy.ca/downloads/AreasOfOperations/EastCoast/DevelopmentApplication/Vol2_Production.pdf

9. Talatori, S. and Barth, T., (2011). Rate of hydrate formation in crude oil/gas/water emulsions with different water cuts. *Journal of Petroleum Science and Engineering,* 80(1), pp. 32-40.

10. Veil. A. J. et al. (2004). "A Whitepaper describing produced water from production of crude oil, natural gas, and coal bed methane". U.S. Department of Energy. [Online]. Available at:
http://www.circleofblue.org/waternews/wp-content/uploads/2010/08/prodwaterpaper1.pdf

11. WS Atkins Consultant Ltd. (2004). Risk implications in site characterization and analysis for offshore engineering and design. *Health & Safety Executive.*

12. Elkins, P., Vanner, R. and Firebrace, J. 2005. "*Decommissioning of offshore oil and gas facilities*". [Online]. Available at:
http://www.psi.org.uk/docs/2005/UKOOA/Decommissioning-Working%20paper.pdf
Accessed 17th December 2013

13. Oil & Gas UK. 2013. "Decommissioning of Pipelines in the North Sea Region". [Online] Available at:
http://www.oilandgasuk.co.uk/cmsfiles/modules/publications/pdfs/OP083.pdf
Accessed 17th December 2013

14. Oil & Gas UK. 2013. "Decommissioning of Steel Piled Jackets in the North Sea Region". [Online] Available at:

http://www.oilandgasuk.co.uk/cmsfiles/modules/publications/pdfs/OP074.pdf

Accessed 17th December 2013

15. Robert Gordon University. 2013. *"Campusmoodle Notes"*. Decommissioning Technology.

16. Osmundsen, P. (2011). Choice of development concept – platform or subsea solution? Implications for recovery factor. [Online]. Available at: http://www1.uis.no/ansatt/odegaard/uis_wps_econ_fin/uis_wps_2011_1_osmundsen.pdf

Bibliography

Fink, J. K., (2012). *"Petroleum Engineer's Guide to Oil Field Chemicals and Fluids"*. Elsevier. [Online] Available at: http://app.knovel.com/hotlink/toc/id:kpPEGOFCF1/petroleum-engineer-s

JAHN, F., COOK, M. and GRAHAM, M. (1998) *"Hydrocarbon Exploration and Production"*. Elsevier. [Online]. Available at: www.knovel.com

Manning, S., F. and Thompson, E, R. (1995). *"Oilfield Processing - Volume Two: Crude Oil"*. PennWell

Appendices

Appendix I

Item	Estimated Costs ($)	Comments
Economics		
Oil Price $/bbl	75	
Gas Price $/000scf	5	
Pipeline		
Subsea – Single phase $mm/km	2,25	20" dia pipe
Subsea – Multiphase $mm/km	1,2	6" dia pipe
Land (/km)	1,25	
Bundle (/km)	5	10" Carrier with four lines
Jackets		
Jack –up – Construct	70	
Jack-up – Rent (/day)	0,35	
Piled	125	
Concrete	150	
Semi-sub	75	
TLP	100	
SBM	25	
Topsides Module ($mm)		
Drilling	75	
Production	125	
Water treatment module	100	
Water injection module	98	
Utilities	65	
Accommodation	80	
Floating ($mm)		
FPSO – Construct	225	150,000 bbl storage
FPSO – Rent (/day)	0,35	
Tender Assisted Drilling TAD (day rate)	0,16	
Shuttle tanker - rent (/day)	0,15	
FSU refurbishment	100	
FSU replacement	150	
Operating Costs		
Fixed (%ge of cum CAPEX)	5%	
Variable $/bbl gross production	2	
Well CAPEX ($mm)		
Sub sea	65	
TLP	45	
Platform	25	

Appendix II

Refurbish (Jacket –Jack up)

Mode Refurbish GAZELLE & FSO including Jack up (for platf wells + water handling + injection)

	REVENUE				CAPEX														OPEX					cash flow	
Year	uction kbopd	Gross product ion	reven ue $mm	cum oil mmb bl	wells	jack et	drillin g	Produc tion	Water Injectio n	Water Treatmen t	FSO	acco m	utilitie s	sb m	pipeli ne	flowl ines	total CAPEX	cum CAPEX	fixed	shutt le tanke r	varia ble	total OPEX	cum OPEX	cash flow $mm	cum cash flow
1	75	88	2053	27		70	75										145	145	7	55	64	127	127	1781	1781
2	76	88	2081	55				125									125	270	14	55	65	133	260	1822	3604
3	74	88	2026	82					98	100							198	468	23	55	64	143	402	1685	5289
4	0	0	0	82	125						100						225	693	35	55	62	152	554	1539	6828
5	70	85	1916	108	125												125	818	41	55	64	160	715	1604	8432
6	69	88	1889	133													0	818	41	55	69	165	880	1751	10183
7	70	95	1916	158													0	818	41	55	68	163	1043	1561	11744
8	63	92	1725	181													0	818	41	55	67	163	1207	1389	13133
9	57	92	1552	202													0	818	41	55	70	166	1372	1231	14364
10	51	96	1397	221													0	818	41	55	77	173	1545	1084	15448
11	46	106	1257	237													0	818	41	55	95	191	1737	940	16389
12	41	131	1132	253													0	818	41	55	0	96	1833	-96	16293

cum oil	cum capex	cum opex	$/bbl
253	818	1833	10.494

Refurbish (Jacket - Steel Piled)

Mode Refurbish GAZELLE & FSO including Steel piled (for platf wells + water handling + injection)

	REVENUE				CAPEX														OPEX					cash flow	
Year	uction kbopd	Gross production	revenue $mm	cum oil mmbbl	wells	jacket-Piled	drilling	Production tion	Water Injection	Water Treatment	FSO	accom	utilities	sb m	pipeline	flowlines	total CAPEX	cum CAPEX	fixed	shuttle tanker	variable	total OPEX	cum OPEX	cash flow $mm	cum cash flow
1	75	88	2053	27		125	75										200	200	10	55	64	129	129	1724	1724
2	76	88	2081	55				125									125	325	16	55	65	136	265	1820	3543
3	74	88	2026	82					98	100							198	523	26	55	64	145	411	1682	5226
4	0	0	0	82	125						100						225	748	37	55	62	155	565	1537	6762
5	70	85	1916	108	125												125	873	44	55	64	163	728	1601	8363
6	69	88	1889	133													0	873	44	55	69	168	896	1748	10111
7	70	95	1916	158													0	873	44	55	68	166	1063	1558	11670
8	63	92	1725	181													0	873	44	55	67	166	1229	1386	13056
9	57	92	1552	202													0	873	44	55	70	169	1397	1228	14284
10	51	96	1397	221													0	873	44	55	77	176	1573	1082	15366
11	46	106	1257	237													0	873	44	55	95	194	1767	938	16303
12	41	131	1132	253													0	873	44	55	0	99	1866	-99	16205

cum oil	cum capex	cum opex	$/bbl
253	873	1866	10.842

Refurbish (Jacket - Concrete)

Mode Refurbish GAZELLE & FSO including Concrete (for platf wells + water handling + injection)

Year	REVENUE				CAPEX													OPEX					cash flow	
	production kbopd	Gross product ion	reven ue $mm	cum oil mmb bl	wells	jacket- Concrete	drillin g	Produc tion	Water Injection	Water Treatment	FSO	acco m m s	utilitie s	sb m	pipeli ne	flowl ines	total CAPEX	cum CAPEX	fixed	shutt le tanke r	varia ble	total OPEX	cum OPEX	cash flow $mm	cum cash flow
1	75	88	2053	27		150	75										225	225	11	55	64	131	131	1697	1697
2	76	88	2081	55				125									125	350	18	55	65	137	268	1818	3516
3	74	88	2026	82					98	100							198	548	27	55	64	147	414	1681	5197
4	0	0	0	82	125						100						225	773	39	55	62	156	570	1535	6732
5	70	85	1916	108	125												125	898	45	55	64	164	735	1600	8332
6	69	88	1889	133													0	898	45	55	69	169	904	1747	10079
7	70	95	1916	158													0	898	45	55	68	167	1071	1557	11636
8	63	92	1725	181													0	898	45	55	67	167	1239	1385	13021
9	57	92	1552	202													0	898	45	55	70	170	1408	1227	14248
10	51	96	1397	221													0	898	45	55	77	177	1585	1080	15328
11	46	106	1257	237													0	898	45	55	95	195	1781	936	16265
12	41	131	1132	253													0	898	45	55	0	100	1881	-100	16165

cum oil	cum capex	cum opex	$/bbl
253	898	1881	11.001

Gazelle replace + pipeline

Mode: Replace GAZELLE complete (Jackets: jack up + concrete)and Pipeline

Year	REVENUE Production kbopd	Gross producti on	reven ue $mm	cum oil mmbbl	CAPEX wells	jacke t	drilli ng	Prod uctio n	Water Injection	Water Treat ment	FSO	acco m	utiliti es	sbm	pipel ine	flow lines	total CAPE X	cum CAPE X	OPEX shuttle fixed	tanker	variable	total OPEX	cum OPEX	cash flow $mm	cum cash flow
1	75	88	2053	27		220	75	125	98	100		80	65				763	763	38	0	64	103	103	1188	1188
2	76	88	2081	55											450		450	1213	61	0	65	125	228	1505	2693
3	74	88	2026	82	125												125	1338	67	0	64	131	359	1770	4462
4	70	85	1916	108													0	1338	67	0	62	129	488	1787	6249
5	69	88	1889	133													0	1338	67	0	69	136	624	1753	8002
6	70	95	1916	158													0	1338	67	0	68	134	759	1782	9784
7	63	92	1725	181	125												125	1463	73	0	67	141	899	1459	11243
8	57	92	1552	202													0	1463	73	0	70	143	1042	1409	12652
9	51	96	1397	221													0	1463	73	0	77	150	1193	1247	13899
10	46	106	1257	237													0	1463	73	0	95	169	1361	1089	14988
11	41	131	1132	253													0	1463	73	0	157	230	1591	901	15889
12	37	215	1018	266													0	1463	73	0	0	73	1664	945	16834

cum oil	cum capex	cum opex	$/bbl
266	1463	1664	11.8

Gazelle replace + FSU

Mode: Replace GAZELLE (Jackets: jack up + steel piled) & FSO (no pipeline)

	REVENUE				CAPEX													OPEX					cash flow	
Year	Production kbopd	Gross production on	revenue $mm	cum oil mmbbl	wells	jacket	drilli ng	Produc tion	Water Inject ion	Water Treatme nt	FSO	acco m	utili ties sbm	pipe line	flowl ines	total CAP EX	cum CAPEX	fixe d	shuttle tanker	variabl e	total OPEX	cum OPEX	cash flow $mm	cum cash flow
1	75	88	2053	27		195		125		100	150	80	65			715	715	36	55	64	155	155	1183	1183
2	76	88	2081	55	83		75		98							256	971	49	55	65	168	323	1656	2839
3	74	88	2026	82												0	971	49	55	62	166	489	1860	4699
4	70	85	1916	108	83											83	1054	53	55	64	172	661	1661	6360
5	69	88	1889	133												0	1054	53	55	69	177	838	1712	8072
6	70	95	1916	158	83											83	1138	57	55	68	179	1018	1654	9725
7	63	92	1725	181												0	1138	57	55	67	179	1197	1545	11271
8	57	92	1552	202												0	1138	57	55	70	182	1379	1370	12641
9	51	96	1397	221												0	1138	57	55	77	189	1568	1208	13849
10	46	106	1257	237												0	1138	57	55	95	207	1775	1050	14899
11	41	131	1132	253												0	1138	57	55	157	269	2044	863	15762
12	37	215	1018	266												0	1138	57	55	0	112	2156	906	16668

cum oil	cum capex	cum opex	$/bbl
266	1138	2156	12.4

Jackets: TLP + wells

Mode Refurbish GAZELLE (including TLP wells + water handling and injection)& FSO

Year	REVENUE Production kbopd	Gross production	revenue $mm	cum oil mmb bl	CAPEX TLP wells	TLP	drilling	Production	Water Injection	Water Treatment	FSO	accoms	utilities	sb m	pipeline	flowlines	total CAPEX	cum CAPEX	OPEX fixed	shuttle tanker	variable	total OPEX	cum OPEX	cash flow $mm	cum cash flow
1	75	88	2053	27		100	75										175	175	9	55	64	128	128	1750	1750
2	76	88	2081	55				125									125	300	15	55	65	135	263	1821	3571
3	74	88	2026	82					98	100							198	498	25	55	64	144	407	1684	5254
4	0	0	0	82	225						100						325	823	41	55	62	158	565	1433	6687
5	70	85	1916	108	225												225	1048	52	55	64	172	737	1492	8179
6	69	88	1889	133													0	1048	52	55	69	177	914	1739	9919
7	70	95	1916	158													0	1048	52	55	68	175	1089	1550	11469
8	63	92	1725	181													0	1048	52	55	67	175	1264	1377	12846
9	57	92	1552	202													0	1048	52	55	70	177	1441	1220	14066
10	51	96	1397	221													0	1048	52	55	77	184	1625	1073	15138
11	46	106	1257	237													0	1048	52	55	95	203	1828	929	16067
12	41	131	1132	253													0	1048	52	55	0	107	1936	-107	15960

cum oil	cum capex	cum opex	$/bbl
253	1048	1936	11.812

Jackets: Semisubmersibles + subsea wells

Mode Refurbish GAZELLE (including Subsea wells + water handling + injection) & FSO

REVENUE

Year	Production kbopd	Gross production	revenue $mm	cum oil mmbbl
1	75	88	2053	27
2	76	88	2081	55
3	74	88	2026	82
4	70	85	1916	108
5	70	85	1916	108
6	69	88	1889	133
7	70	95	1916	158
8	63	92	1725	181
9	57	92	1552	202
10	51	96	1397	221
11	46	106	1257	237
12	41	131	1132	253

CAPEX

Year	Semi-wells	Semi-submers	drilling	Production	Water Injection	Water Treatment	FSO	accom	utilities	sbm	pipeline	flowlines	total CAPEX	cum CAPEX
1			75	75									150	150
2				125									125	275
3					98	100							198	473
4	325						100						425	898
5	325												325	1223
6													0	1223
7													0	1223
8													0	1223
9													0	1223
10													0	1223
11													0	1223
12													0	1223

OPEX and **cash flow**

Year	fixed	shuttle tanker	variable	total OPEX	cum OPEX	cash flow $mm	cum cash flow
1	8	55	64	127	127	1776	1776
2	14	55	65	133	260	1822	3598
3	24	55	64	143	403	1685	5283
4	45	55	62	162	565	1329	6612
5	61	55	64	181	746	1383	7996
6	61	55	69	186	931	1731	9726
7	61	55	68	184	1115	1541	11267
8	61	55	67	184	1299	1369	12636
9	61	55	70	186	1485	1211	13847
10	61	55	77	193	1678	1064	14911
11	61	55	95	212	1889	920	15831
12	61	55	0	116	2006	-116	15715

cum oil	cum capex	cum opex	$/bbl
253	1223	2006	12.782

Appendix III

Table – 7: Wellhead alternative

Wellhead Platform					
Jackets	**Average fabrication time**	**Cost $/bbl**	**Advantages**	**Disadvantages**	**Comments**
Jack up	24 months	10.5	1. Steady and relatively motion free platform	1. Weather dependent for placement	Wellhead CAPEX used as a basis for evaluation
			2. Low cost and efficient. Used in shallow water up to 100m such as the Gazelle field water depth	2. Seafloor scour	
			3. Drilling package can be uploaded with higher hook loads and installation of top drives	3. Blowout can cause collapse of platform due to soil fluidization	
Piled	24-30 months	10.84	1. Support large deck loads	1. Costs proportional with depth	Wellhead CAPEX used as a basis for evaluation
			2. Can be constructed and transported to site	2. Maintenance costs very high	
			3. Piles provide good stability	3. Not reusable, steel structure members subjected to corrosion	
Concrete	55-60 months	11.001	1. Support large deck loads	1. Cost proportional to depth	Wellhead CAPEX used as a basis for evaluation
			2. Construction can be completed before floating and towing to site	2. Requires more steel than jacketed platforms	
			3. More tolerant to overloading and sea water exposure	3. Foundation settlement subjected to seafloor scour	

Table – 8: TLP alternative

TLP Wellhead

Jackets	Average fabrication time	Cost $/bbl	Advantages	Disadvantages	Comments
TLP	24-30 months	11.812	1. Improved motion characteristics	1. High initial subsea cost	Wells TLP CAPEX used as a basis for evaluation
			2. Possibility for full integration and commissioning prior to installation	2. Payload limited	
			3. It improves drilling and workover capabilities (Surface wellheads)	3. Vertical mooring system does not provide control of horizontal position (e.g. for well access)	

Table – 9: Semisubmersible alternative

Subsea wells

Jackets	Average fabrication time	Cost $/bbl	Advantages	Disadvantages	Comments
Semisubmersible	36-48 months	12.782	1. Less sensitive to water depths	1. Expensive to operate	Subsea wells CAPEX used as a basis for evaluation
			2. High load capacity	2. Expensive to build	
			3. Installed as a full integrated system	3. Drilling and workover capability will have to be through subsea BOP